AF323182

ANY
BODY'S
DAY

John Frisby

# Anybody's Day

Blond and Briggs

First published 1978 by Blond & Briggs Ltd
London, and Tiptree, Colchester, Essex
© Copyright 1978 by John Frisby
SBN 85634 050 2
Printed in Great Britain by The Anchor Press Ltd
and bound by Wm Brendon & Son Ltd
both of Tiptree, Essex

for Sheila

# Acknowledgements

To Professor John Yudkin
for his advice and encouragement,
and Eduardo Paolozzi for his enthusiasm –
thank you.

Every theory of the course of events in nature
is necessarily based on some process of simplification
of the phenomena and is to some extent therefore a
fairy tale
*Sir Napier Shaw*

What is now proved was once only imagined
*William Blake*

ANY
BODY'S
DAY

The machine lay on its side.

Beneath its surface, complex motors and servo systems
carried out their automatic repairs to damaged or
over-stressed circuits and components in almost total
silence. The majority gradually returned to near-peak
performance. Some, however, continued to
malfunction, and a few circuits showed the first distinct
signs of eventual breakdown; making replacement
necessary for the machine to continue to operate.

Its two irreplaceable linked computers were housed in
the upper and lower levels of the machine's control
centre. They were made of soft grey jelly and were
almost ninety per cent water, electronic water,
weighing no more than a can of Coke.

Routine electrical wave patterns washed through both
computers, recording and monitoring the circuits'
activities with each swell. The smooth idling rhythm of
ten cycles per second was synchronised by tiny
pacemakers in the lower computer. In the furthest
circuitry of the upper half they probed the shuttered
and temporarily undisturbed mechanics of the
machine's intelligence. The computers were a

biological clock, regulated by the Sun and the Moon.

While animals slept within their bodies, their faces
waking first, the machine woke before his face.
He slept with his body in his mind.

The casually soothing wave tremors were destroyed by
minute changes in electrical pressure. The pacemakers
automatically sought to cope with the new irregular
wave patterns that fed disjointed signals throughout
his brain's open circuit fibres. Millions of the
microscopic wave impulses, electrically charged
sodium and potassium, were boosted 1000 at a time
at 100 metres per second.

The effect on the machine's communication system
was instantaneous.

Disjointed fragments from the stored memory tapes
flooded the motivation circuitry in the upper computer
with disturbing fantasies. Both eyes started to flicker
behind their closed eyelids. Their blind activity
instinctively struggled to rationalise their memories by
visual reassurance, as the machine's automatic
warning system tripped its relays to defend itself from
the overload of autistic distortion.

*I blinked.*

Hearing – the machine's distant sense of touch –
picked up the first detectable sound waves.
The arriving vibrations contained an invisible jumble of
intensities and frequencies of sound not yet
individually identified as traffic, a clock ticking, and the
rhythmic breathing from a similar adjacent machine.

The apparently careless design of the two squashed-up
audio-receivers, fixed to the left and right of the
machine's control centre, prevented any sustained
build-up and emphasis of specific sounds. A
microphone within each receiver picked up the sound
waves that penetrated its soft protective film of
yellow wax and converted them into micro-electrical
impulses. Each ear was so visually different that it
could be used to distinguish the machine from any
other. In conjunction with the computers, they were
concerned more with the location of individual
vibrations. This selective ability heard what it wished
to hear – without selection, the machine's
surroundings were impossibly noisy and raucous.
The two receivers formed a stereo-system that enabled
the computers to identify the source and to tell
whether it was moving. The impulses took a
700-thousandth of a second to travel along
message-fibres to both computers, checking and
interpreting the disturbances produced by the
vibrating objects, comparing them with the memories
of other mornings. The sensitivity of the system came
close to the theoretical limits imposed by the basic
structure of matter. The sensation of sound which
could be picked up and recorded by a movement of
either microphone was a thousand-millionth of a
centimetre, and within the microphone the movement
of a further refining membrane was a hundred times

smaller – too small to be seen by a microscope. His ears also housed the vital navigational control and stabilising mechanisms – fundamental to the machine's balance.

Beneath the computers on the front face of the control centre were twin oval shutters, mounted a hand's width apart. Silently, they opened for a tenth of a second, then an eighth, two seconds, a fifth – then closed again.

Behind the shutters, two cameras – the size and weight of marbles – recorded the limited visible section of the machine's immediate surroundings. Once through each tiny lens, with its 2000 transparent layers of tissue and distinctive dark brown outer ring, the fragmented, partially focussed, chromatic pictures flickered through a tiny, clear fluid-filled chamber before striking the rear sensitised projection wall. To develop the pictures, the wall was studded with more than 130 million microscopic rods and cones – light and colour sensitive cells – linked by message-fibres connected directly to both computers.

The curved wall had been built as a miniature extension of the computers. Together, the two walls daily processed four billion chromatic binocular images. The first 200 pictures that arrived for processing were mostly of a blurred sideways view of a small circular dial, an arm's length away. Consecutively spaced around the dial's perimeter were the numerals one to twelve. Two pointed metal strips of contrasting length radiated out from the centre, moving clockwise at differing speeds.

*Christ, what's the time?*

Collectively, the numerals subtended an angle of five minutes, or one-twelfth of a degree, from the cameras. To be processed, each had to be shrunk down to an *inverted* image almost too small to be seen. It was this extent of reduction and complexity of processing that the computers and communications network had to decipher if they were to believe in the machine's visual competence. Consequently, it was of no help that the message-fibres in his optic nerves had become softened during the previous eight hours of inactivity. Softening inhibited transmission, overloaded individual signal connections, and temporarily fused any visual communication.

*Okay for another ten minutes ... another Thursday. Another morning to wait for a second or two before remembering why I had to get up and go to work.*

He relaxed. Any breakdown in visual communications, however temporary, automatically activated the remaining warning systems in the lower computer. Despite his conscious indulgence, the machine's increased electrical wave patterns opened memory circuits, linking in cross-feed from the analytical circuits which immediately requested ordered sequential information.

The machine's essential advantage over all other types of machinery was not due to the possession of bigger computers, but to the structure of the interconnections between their circuits. The lower and older computer had taken 300 million years to develop and now contained the systems essential for day-to-day maintenance, overall management, navigation, stabilisation, external and internal communications, and other coordinating functions. It was programmed to operate automatically, largely out of his control. As the machine's tenant, he chose to ignore it and negligently accepted every decision it took.

He preferred to inhabit the upper and larger computer - a relatively recent innovation developed over the past half-million years. Through the experience gained from countless previous models the upper computer contained memory, logic and emotional circuitry, and was now capable of enjoying to the full the most exciting of its inventions — abstract thought.

The condition of the machine's computers, after only thirty-six years, should, however, have been of major concern to the tenant. Two-thirds of their potential circuitry — including telepathy, orientation, a

knowledge of time and the ability to predict
earthquakes — had never been used and, due to his
indifference, self-indulgence and lack of use, they
never would be.

The memory circuits were hardest hit. Many of the
millions of tiny complete memory packets had neither
been stored nor transferred cleanly. With each recall,
they left behind a small fraction of the packet. The
original clearly defined memory prints had gradually
become smeared with time and use. In a long recall
the signals were becoming meaningless, especially
as poor change-transfer increased.

His memory capacity and flexibility were steadily
diminishing by several thousand cells per day that
doubled whenever he was drunk. Although
8000 million remained and collectively continued to
operate, the computer's activities increasingly relied
upon a bank of circuits coded 'Routine'. Now
dominant, this in turn chose to connive with his
ill-equipped memory in obtaining access to the entire
circuitry of the combined computers during his
passive periods which occupied a third of each day,
spent in the cyclical depths of sleep.

500 thousand years earlier, primitive models of the
present lower computer had invented an ingenious
biochemical trigger to regularly irritate earlier tenants
out of their dormant lives. This imbalance had
developed as their machines changed from low-grade
vegetable fuels to high energy-producing meat fuels,
moving faster to avoid extinction and to operate
almost continually during the hours of daylight.

Obtaining fuel was difficult and dangerous at night. The essential amino-acid trigger was only needed for the approximate sixteen hours of daylight followed by eight nocturnal hours when the necessary auto-mechanical repairs could then be carried out in relative safety to the inactive and defenceless machine. The relay in the lower computer that controlled the trigger moved to 'go'.

Part of the machine began to move.
Above his body, an invisible column of atmospheric pressure eight kilometres high contained within the shape of his outline and weighing more than half the weight of a car, attempted to flatten him. The machine automatically resisted with its own comparatively minute quantity of pressurised air dissolved in its components and circuitry. His skin was the dividing line between these two pressures which, as long as they remained in balance, permitted them both to survive.

Immediately below the control centre on each side of the machine, thirty-two of its 206 structural components gently articulated, hinging away from the main body of the machine like the exploring jib of a crane.

Each component was different in shape, weight and size. Gram for gram they were stronger than steel. The joints between them were self-lubricating, capable of operating for more than seventy years without servicing.

To merely stiffen his arms, enabling them to translate his instructions into a wide range of delicate pressures

and violent forces and power, were fifty-eight
muscular motors – thin fibres, little more than
striped jelly.

The muscles acted in pairs, one opposing the other.

Signals from the computers adjusted the tensions in
each muscle so that their energies did not exactly
cancel out, but resulted in the assembly's smooth
controlled movement.

The ends of each arm first broadened and then
divided into five jointed members for probing and
gripping. One of these, longer and fatter than the
others, had played a major role in the machine's
evolution. Within the computers there were more
circuits involved in the manipulation of his thumbs
than there were controlling the machine's major pumps
and refuelling systems together.

The ends were cushioned on the under-side with a
unique oval pattern of ridges, terminals, spurs,
crossovers, divisions, enclosures and islands to
improve grip and distinguish the machine from any
other. No fingerprint was exactly alike, and each
remained in constant pattern throughout his life.
The tenant kept them carefully shaded as too much
sunlight made his fingerprints disappear. On top were
hard, flat, transparent, horny shells to protect the ends
and form additional weapons. Their roots were buried
in grooves in the machine's skin, growing continually
during its life and for months after the tenant's death.
Of the tenant's five million sensory fibres, one-third of
them were to be found on his fingers, thumbs and the
palms of his hands.

His hands were the principal instruments on Earth.
The first tools. The whole of his technology stemmed
from these creative, manipulative components.

They were covered in the machine's tough, self-
sealing skin — more versatile than any other known
engineering material — protecting and monitoring its
external surroundings with its buried layers of contact
and temperature sensors.

The energy that the Sun was radiating on Thursday
had taken 50 million years to percolate the 600,000
kilometres from its core to its surface.

Travelling at the speed of light, it then soundlessly
blasted out its latest, raw, life-destroying radioactive
energy at the solar system. Eight and a half minutes
later it reached Earth. This enormous separating
distance would not have prevented the continual
violent solar flood from turning the Earth into a deadly
wasteland like the Moon. Instead, wobbling along at a
vulnerably slow twenty-seven kilometres a second in
its first uneasy spinning orbits around the Sun, the
Earth had constructed its own ingenious defence.

Using gravity and the huge electro-magnet in its core,
it had built a protective filter — the paper-thin

biosphere – quickly refining it into a substantial heat pump running on the invading solar energy.
The pump stirred round a dense, insulating, slowly churning soup. The soup contained water vapour combined with gases, and weighed more than all the human machinery on Earth.

As light as air.

To prevent itself sticking to the Earth, the soup developed a permanent, minute clarified layer, a bare kilometre deep along the Earth's rough surface.

The machine could only operate unaided if the Earth's surface kept within this layer.

Through the biosphere, Thursday's sunlight overtook the disfigured outline of Turkey, then Central Europe, first picking out the Alps before reaching down to Greece and Italy, travelling West at over 1500 kilometres an hour as the Earth continued on its latest invigorating spin.

Three minutes later it reached the machine.

Deep inside his body, the previous eight hours of inner darkness instantly changed, revealing the silhouette of the machine's structure to the deeper, more delicate components and gently pulsating communications networks. The shafts of sunlight disclosed the innermost complex engineering, finished in brilliant rose-pink, scarlet, yellow and mauve.

*I shivered, pulled the blankets away from my back and, without waking her, sat up.*

A hot Thursday morning. At the Equator, the blanketing
heat haze lifted slowly over Antarctica, joined at the
South with the blurred outlines of the British Isles and
North-West Europe: one corner of the connected
jigsaw puzzle of all land on Earth. 500 million years
before Christ.

In the warm shallow seas rode millions of the machine's
tiny sightless ancestors, feeding on the delicate
organisms produced by the effect of sunlight on the
upper soup-like layers. They quickly developed
sensory cells to guide them up unerringly from the
protection of the seabed to their sunlit feeding grounds;
sinking back once more at dusk. To survive as a
particular type of machine for just 100 million years,
they needed more sophisticated extensions to a
communications system that was solely geared to
moving slowly up and down. To grow in competition,
they needed better methods of warning, navigation
and external communication. To coordinate these they
needed a computer. At first it was merely a swelling at
the front end of the sensory cord – no more than a tiny
blister – then another and another: the base on which an
eventual computer would be built. For a long period in
the machine's evolution these earliest chambers were a
sensory system in themselves, expanding to contain
fear and curiosity circuits; driving and inducing the
developing explorers to follow their plant food which
had already begun to grow on land.

To survive ashore, enormous changes had to take place;
refinements, accidents, evolution within evolution –
300 million years before the machine's cameras were
fully developed from a *sense* of vision. Their built-in
receivers that had been so sensitive to vibrations

through solid water finally became outdated by an extraordinary stereo sound system.

While the second computer was being developed by earlier machines, their tenants communicated by telepathy. They had no fire, no tools except their thumbs, no weapons but rocks. There were poisonous plants everywhere. Storms and freezing cold. There was no speech; only grunts. Yet there was a form of communication to tell each other which plants were edible or poisonous, or which animal tracks to follow or not. A telepathic system existed, linking the computers of plants and animals. Each crude machine immediately knew what was stored in neighbouring memory banks. They sensed each other's thoughts.

There was a long-term exclusive dependence upon telepathy as a method of communication, which continued even after the development of the new computer when reasoning power became more advanced. It was used for millions of years more, finally giving way to other methods of communication that the machine and tenant now jointly relied upon. The residual circuits, however, still remained. Electrical wave patterns flowing through the computers invariably swamped the now undetected telepathy stored in the middle capsule in the base of the old computer. In his infinite capacity for evasion, the tenant had carefully printed special computer circuits containing fantasy and illusion. His own romantic compositions had already prepared the ground for his enjoyment of totalitarian politics. Both dealt in dreams, and both knew that the illusions the tenant had about himself as an individual were more persuasive than reality.

*I sat on the edge of the bed and looked down at my cock.*

A microscopic world of 140 million parasites was located in the surrounding dense and humid jungle. If he did not scratch with just one of his fingers or wash this world for a week, it would give the parasites an extra 30 million years in their comparative time.

The computers stared at themselves, appearing to each other like strange gardeners in an alien world where seeds that they invented could germinate just by thinking about them. Fear and envy, fantasy and illusion: these were the plants they tended and their tangled roots spread over into the tenant's conscious thoughts. If he could not or chose not to reactivate the residual circuits of communication, then his computers would eventually rid themselves of their terrors by destroying them both.

*I yawned, stretched and started to stand . . .*

The computers introduced and coordinated thousands
of instructions to carry out these three simple
movements.

Like shellfish opening and closing in time with the
ocean tides, catching food and circulating internal
water and refuse, the machine had clung to the same
mechanisms as its tiny ancestors. When they had
finally decided to remain on dry land they took their
ocean and tides with them. The machine was sixty
per cent water, a vertical puddle with a mineral
content almost exactly that of the oceans 500 million
years ago. On land the safest and most convenient bed
for its own blood-red ocean was inside itself.
Its shoreline was the soft pink undulating inner
surface of its own skin. Its tides were now internal.

The machine had been rhythmically taking in air
twelve times a minute and although it only used a
fifth of the oxygen content, its two sterile porous
bellows traded clean moist air for saturated old with
the surrounding delicately overlapping deltas where
the internal tides constantly ebbed and flowed.

The machine's pear-shaped tidal pump effortlessly
circulated the ocean, using it as a fluid assembly line,
delivery network, and waste disposal system; all
regulating the machine's internal temperatures.
Each year it lifted the equivalent of thirty-nine
B-52 bombers — and their bombs.

It was with this enormous strength that the exchange
of fuel for waste was continually sought.

Within the changing ocean, the machine's primary

defence mechanisms against infection kept vast armadas continually at war. One salty red drop contained more than 250 million defence cells.

The machine had already been nearly defeated this morning. The deadly, uncontrolled growth of the now extinct cancerous cells had been swept away on the tide, filtering down the long central disposal system.

Throughout the day the properties of the ocean went through huge cyclic changes. After eight hours of rest and repair the machine's mobile efficiency had dropped to a condition which the lower computer's monitoring systems could no longer tolerate. The tenant's idle yawn and stretch brought to the machine its requested bursts of energy to tension essential structural joints and revitalise sluggish components.

Both computers high in the central control were smoothly lifted upright. The sitting machine hesitated, the computers getting through that detailed checklist which so long ago had separated the machine from all others.

180 structural components stiffened and then, commanded by 270,000 carefully spaced signals barely lasting three seconds, the machine began to pull itself to its maximum height.

*I held my knees as I began to stand . . .*

Here the lever forces soared . . . smoothly, swiftly, up
and up until the forces at each joint reached six times
the weight of the machine, before gently descending
back to zero, the entire structure balancing unsteadily
on two feet.

*'Hello.*

*I turned at my wife's sleepy v oice and bent down and kissed her good morning –*

— Unaware as he did so that he was gently drawing on a nine-metre tube, and that the end of it was full of shit.

Now at his full height, the tenant did not see things as they were, but as he was.

The vertical machine began to walk. Over fifty muscular motors tightened or loosened the selected connections for the unstable, two-legged structure to subtly change shape and position without the need for elaborate shock-absorbers. To remain healthy, his bones had to be continually under stress.
Stress produced surges of electrical current throughout their smallest fibres as the machine adjusted the length and inclination of each moving part. Cantilevers instantly changed into columns, beams into buttresses, as the current flowed from positive to negative and back again. The electrical surges stimulated the fibres' cells to pump the fuel supply through the hollow structure's narrowest channels. Electrical fractions of time in motion were directed by the impressive muscular power system that needed no rotary movement. The wheel, that basic invention essential to almost every form of transport, did not exist in nature.

One step after another. The fact that one and one apparently made two was assumed by the tenant to mean that his shift of circumstance was unimportant. It seemed impossible for him to analyse this degree of unimportant change. In complete secrecy, the computers had not merely increased the size of the machine over the years to its present height, but entirely *replaced* the rest of the machine on two occasions. Apart from the computers, the machine that the tenant now occupied did not exist ten years ago. Each change had taken longer than the last, and after the next there would be no more. And now the machine had begun to shrink. Without help, the complex mechanical components were irreversibly running down.

Motors in the machine's legs that had become stiff with lack of fuel during the night caused the machine to momentarily stagger, shaking its components — especially the waste disposal mechanisms.

Buried in the body of the machine, slightly to the rear and either side of its main supporting column were two duplicated chemical pumps. Duplicated in case of partial or complete failure. Hardly bigger than hens' eggs, each kidney contained a quarter of a million transparent filter pipes, more than fifty kilometres long. They were some of the machine's most complex mechanical components; delicate monitors of more than thirty of the internal ocean's essential chemicals. Removing some, conserving some, and delicately adjusting the levels of others, they maintained the sea's essential acidity and stabilised the pressure and volume of the tides.

The filtered extract was a useful pale yellow liquid, continually dribbling out of the bottom of the filters through two narrow pipes down to a waste disposal bag the size of a beer glass. The rate of flow had slowed to a quarter of its normal output during the night, but now the bag was almost full. Clinging to its walls were finely adjusted sensors, generating signals to the computers that corresponded in urgency to the amount of increased pressure. Beneath the bag were two release valves. As the pressure rose, the machine's computers automatically instructed the upper valve to open, making the tenant conscious of his temporary control of the lower.

*I needed a pee.*

The curved levers clamped around the top of the bag contracted, then those ringed below added their compression, wringing the bag out, forcing the liquid down under pressure and collapsing it into rhythmic folds.

The torrent pumped down the exit pipe, the size of a pencil lead cunningly terminating inside a handy, flexible tube.

*I squeezed.*

From a lovingly assembled congestion of pipe fittings
– delicious relief. Hot piss. In its fresh state, almost
sterile. To the injured in battle, an emergency
disinfectant to be dispensed straight on to open
wounds. The planned hydro-dynamics in the twisted
spiral of the masculine stream ensured the hygienic
compact flow to the wounded – or the china bowl.

*Aah.*

*My tongue felt the precise bevel where my teeth
rested against each other.*

The machine's touch receptors responded to the size
of the toothbrush and its weight, length, balance,
temperature and texture. They translated these into
minute chemical signals – three-dimensional electrical
codes – and transmitted them to the computers.
The signals were stacked in precise relation to other
sensations simultaneously received from the
machine's impressions of sights, sounds and smells.
The construction completed, it was carefully checked
against the blueprints of past toothbrush events
stored in the memory banks.

Glimpses of twenty-eight stunted enamelled trees
appeared in the mirror. The last remaining ivory-
coloured replacements for the first twenty deciduous
stalks had been inadequately developed for a life of
barely forty years. Together with his tonsils and
appendix, they were the only sections of the machine
that had not kept pace with evolution. To continue to
fit his mouth they had to be regularly ground down.

The two semi-circular rows easily decayed.
The combined effect of toothbrush and toothpaste
was less efficient at cleaning his teeth than scrubbing
with a twig from a tree. No other structure in the
machine was as likely to perish as his teeth, yet no
other structure was as resistant to decay after death.
Castration was the sole remedy for baldness and
acne, as death was the cure for cavities.

*I turned on the cold tap and rinsed out my mouth.*

Four days earlier, the recycled water that was now flowing over his gums had been a cup of tea sipped by a seventy-two-year-old lady, ninety kilometres away. During the next four days, the tea had been pumped underground, joined a river on three occasions and been chemically processed eight times. Between processes and trips to the river it had been drunk with Alka-Seltzer to cure a hangover, then by a baby forty kilometres nearer, followed by a pregnant girl who had swallowed all three only yesterday. Finally, clear again, into his mouth. A taste as old as cold water. Ten hours more, and it would help to carry an oil tanker out to sea.

The machine's balance of internal heat and waste processes kept a continual minute flow of moisture seeping through its skin. A tiny gland had overflowed, blocking its exit, and as it leaked between the adjoining layers of skin it had erupted into a white rubbery cone. The skin's normal bacteria thrived on the trapped and concentrated waste.

*I squeezed the spot on my chin.*

With even the smallest damage to its surface, the machine's defences signalled the adjoining cells to surround the cone with an inflammatory ring by increasing the temperature and blood-flow in the area. The blood circulation's smallest branches and openings widened, letting through the clear fluid defence cells to eat or neutralise the remaining infected waste and damaged skin around the puncture. They then began the repair by weaving a criss-cross patch with cell debris. The entire ring was walled off and isolated by a temporary block in circulation and anything that escaped was killed off in the machine's deeper strategically placed defence components.

Squeezing the spot produced a short sharp pain. After a perceptible delay it was followed by a dull, disagreeable and longer-lasting ache. The severed fibres surrounding the puncture sent repeated alarms as pain signals to the computers. Each sensation did not add to another, and was no stronger than the last. Deeper pain, however, had a higher threshold, building up as the stimulus and its effect accumulated over a period of time, carrying the closely connected language of discomfort either separately or simultaneously. The language was so precise that the machine could distinguish more than twenty levels of increase in pain, from the superficial puncture to the impending failure of its deepest components. The messages were carried by an intricate network of a quarter of a million separately insulated fibres grouped in varying thicknesses. Faster signals travelled in the thicker groups, slowing in the thinnest fibres at the computers' terminals.

The computers were the only part of the machine

unaffected by pain. As there were no pain receptors, whole sections could be cut out, crushed or melted, but not rebuilt. Their distress was the idea of pain.

The potential hidden dangers in the machine's internal and external worlds were constantly checked and compared, as the computers assessed the information from the tenant's awareness. 100 million impulses of analysed information passed to and from the machine's central nervous system every second.

This mass of information, often contradictory, would be overwhelming if the machine instantly responded to each message. Instead, the messages passed through a filter system in the lower computer, to enable a fraction of the information to be processed piece by piece depending upon its priority. Most of the decisions were then automatically dealt with before they could reach the upper computer and be consciously interpreted by the tenant. Each impulse entering the network was faced with 10,000 million interconnecting on/off switches permitting or preventing a message going any further or rerouting part of it or all of it, or even duplicating it for separate transmission. In an emergency, a message could be instantly increased in speed from 200 to 300 kilometres an hour, to travel two metres.

The noise of the electric shaver was a sudden stench in the ear. Noise was now the chief product and authenticating sign of evolution in technology.

*I switched off the irritation and stared at my reflection . . .*

The tenant had only seen himself left to right and had always imagined that he looked like his reflection, but no face was remotely symmetrical. The noise . . . could his ear really smell it? Was it possible to 'smell' noise or 'taste' colours? Could he taste colours never seen before, smell sounds never previously heard? The senses would become tangled; the wrong messages delivered through networks intended for other uses, finally 'hearing' the colours and 'seeing' the sounds, questioning his basic structure of reason.

The shaver buzzed on. Only a few areas of the machine's humid surface were completely visible. Its nine openings and the special supporting and gripping surfaces were exposed; the rest was covered in a varied growth of furry protective cells, supplying the machine with a sensitive awning of touch. Each hair cell was a brown wavy hollow tube delivering scented oil from within the machine's tissue-paper-thin wrapper, to slowly build up a waterproof coating over its skin. The most concentrated section of the awning was a canopy of over 100,000 hairs covering his head, insulating the computers from extremes of heat and cold; cushioning his head against potential dents and gashes. Each hair lasted between two and five years, and there were nearly 100 shed each day. The tenant combed back his hair, raking up the tiny discarded cornflakes of dead cells from his scalp. Up to a quarter of a million bacteria were embedded in each flake of dandruff.

*As I brushed the dandruff from my shoulders I dropped the comb, but caught it before it hit the basin.*

Whilst the comb was in the air, the stationary machine
had moved forward in space-time with the spin of the
Earth. When the comb rejoined him, his hand had
moved forward, and the comb had also jumped
forward to keep pace with it, taking a quick corkscrew
journey through space.

Beneath the canopy were two additional curved
protective strips of hair, one over each camera, to
prevent water flowing uninterrupted down the face of
the control centre and into the unshuttered lenses.
The computers' protective shell was a jointed helmet
of twenty-two irregularly shaped armour plates.
Fourteen of them were hollowed and lined with a
liquid shock-absorber to take the stacked computers,
the twin cameras, the navigational controls and
stereo-system. The front lower edge was a hinged
filler cap, concealing the machine's long digestive
tube. The entire structure of the control centre had
developed logically around the tube's entrance,
helping it to seek out and test possible sources and
grades of fuel.

The fastest growing hair surrounded the filler cap.
It had been mown flat each day for the past twenty
years. When a moustache and beard surrounded his
lips, the fleshy edges became bright red, turning his
mouth into a cunt fringed with bushy pubic hair – an
unexpected sexual choice.

The oily brown shavings from its 40,000 hairs rose in
an imperceptible dust cloud, drifting into his eyes and
nose and mouth, sticking to the moist surface of the
surrounding skin.

Hot water first. Usually too hot. Almost half the
water's temperature was quickly lost in heating up the
china basin. The temperature-checking overheated
fingertips reactivated their pain signals in the
computers, rousing the last sluggish circuits of
awareness, then stimulated him with the small
comforting shocks of splashed cold water in the face.
Water was the best solvent in existence.
It was extremely stable and could be easily frozen or
vaporised, carrying many chemicals either in suspension
or solution, without itself being changed. The warm
water was now slightly above the machine's surface
temperature. Two minutes' contact was all that was
needed to soften the skin and hair and float off any
unwanted layers without damaging its surface.
The soapy water produced an artificial softness,
completely removing the machine's waterproof
coating. Washed and dried, the tenant felt better.
The machine, however, would take nearly an hour to
retrieve its natural covering, wasting energy that the
tenant needed and instead would now have to get by
refuelling.

*I walked into the kitchen.*

The deep window framed the spring-green leaves of
trees – large vegetables almost entirely made of wood.
Trees lived in a leisurely way, refuelling slowly from
the air, water, sunlight and soil without any exertion,
taking exercise whenever the wind blew.
Like the machine, if the branch of a tree was loaded
too long, the wood gradually ran away from the
weight; the fibres taking advantage of the natural
changes in moisture and temperature to shuffle
away from the local excess of gravity.

The smell of fresh coffee stimulated his thoughts of
breakfast. Orange juice first. The volume of dilute
acids indicated the machine's instinctive craving for
acid fuels. The urgency had lasted sixty-five million
years following the sudden reversal of the Earth's
magnetic field, North Pole with South. During the
thousand-year switchover, its protective outer layers
had fallen to almost nothing. The ozone filter had
quickly become exhausted, causing a lethal
downpour of ultra-violet radiation. The effect was
widespread and devastating. Without warning,
plants – edible for millions of years – suddenly
became poisonous, flowering into alarming Gardens
of Eden full of unknown anaesthetics, toxins and
hallucinogenics. This catastrophic reversal played a
dominant role when one-third of all species of life
died out.

The surviving tenants carried mutations in their
blood that speeded up their evolutionary change.
Crucial components and connections in their
machines had been destroyed or altered beyond
repair. Equipment that had previously assembled
a specialised acid – vitamin C – could now only be

obtained from the increasing number of readymade
fuels. However, as they no longer used their energy to
operate and maintain their former vitamin-C-
producing equipment, they had a clear advantage over
other animals which still had their traditional
equipment intact. The machine's acid fuel now
included both plants and animals. It could be powered
by almost anything organic.

The details of the complex mutations were filed in the
microscopic blood cells that coloured the machine's
salty oceans red. Five million in a pin-prick.
Twenty-five billion in constant circulation. Each one a
specialised chemical processing plant carrying
dissolved air to the smallest part of the machine in
continual supply. Although the machine could
operate for several days without refuelling, it could
only last a few minutes without air.

As fresh air was sucked into the machine, the
revitalised discs turned the tides bright red. Without air
the oceans would be thick, black and sticky – crude
oil – unable to penetrate the intricate engineering.
The sluggish recirculated currents returned to his
lungs dark blue with little air left in them, then were
urgently refined back to purple, then dark red; finally
retrieving the most effective lubricant known.

Each disc could survive 170,000 circuits of the
machine before wearing out and needing replacement.
None of them lasted longer than four months. Their
individual structure was a precise web of 574
amino-acids. Only one mutation separated it from that
of a gorilla. A change could be expected every ten
million years, and barely one-fifth of a mutation

divided the present machinery from its cave man
ancestor. However, the changes were indivisible.
Either the complete separation had already taken place
or it was coming now.

*I sniffed the orange juice and gulped it down.*

The sharp intake of air through the divided ventilation grilles overhanging the filler cap broke up the normal smooth air flow, sending swirling smell-laden air high up into the machine's coupled air-conditioning units before being sucked down into the porous bellows.

The tenant's nose with its half-hidden tubes were combined together into the finest air-conditioning plant ever made. They cleaned, warmed and moistened the incoming air to protect the bellows' spongy linings, ensuring the best possible vapour-exchange in their sterile networks.

Embedded in the ceiling of each nostril was a brown strip, the size of a postage stamp. The strip was printed with fifty million floating smell receptors, linked to a tiny local computer for initial data processing before transmission through message fibres to the central control.

The messages were precise and accurate. Once stored in the main computers, a smell was almost unforgettable. The tenant could distinguish thousands of different odours, but, like attempting to describe colour to a blind man, he could not explain the experience of a single new aroma.

The machine's fuel supply had been initially restricted by the tenant's senses of smell and taste. Taste had a very short range, reacting only to fuel already inside the machine. To increase the range the machine had introduced touch, colour and appearance into the selection. Without the last two, the tenant could not tell the difference between an orange and a grapefruit.

He felt hungry, and although he could hardly believe
it, his stomach never lied. Eating was touch carried to
the bitter end. Although nature used as little as
possible of anything, never breaking its own laws,
there was nothing manufactured that was not without
short cuts. The acid chemicals in the hot black coffee
stimulated the computers' circuits, increasing the
tenant's capacity for work as it decreased his
persistent feeling of being only half-awake.
The expanded network connections compelled his
thought processes to flow more quickly, although not
always in coordinated sequence.

The filler cap reopened. Inside the tilted funnel a
sample of the likely fuel received an on-line analysis
from his tongue in a four-hundredth of a second.
More capable than any laboratory. If the fuel tasted
good, the tenant swallowed it.

Yet there were only four tastes – sweet, salty, sour and
bitter. These were symmetrically arranged around the
perimeter of the muscular, oval tongue firmly anchored
at the back beneath the entrance to the digestive
tube, redirecting the air flowing in and out of the
machine, modifying the tenant's voice, formulating
his speech.

Its healthy pink surface was concealed by a soft white
furry coating, formed overnight from gases given off in
reaction to an excess of alcohol in his stomach that
had combined with his tongue's reaction to the
nicotine and tar from too many cigarettes. It had a fine
pattern of 9000 closely packed sensors, each with its
own tiny protective pink umbrella – merging together
communicate a consensus of opinion.

Honey. A mouthful contained the visits to 5000 flowers, 200,000 kilometres per jar.

For the sensors to work, the fuel had to be at least partially liquid. Dry fuel had to be moistened before its taste could be detected.

From beneath his tongue, more than a litre of lubricating water flowed each day, starting off the fuel's chemical breakdown, although a quarter of it was used in making speech possible, keeping every part wet and ready.

Hard fuel was accepted on approval, torn off into small pieces, and then dished from side to side to the enamelled grinders. The crushed pieces were then retested. Anything soft was pushed against the roof of his mouth, pressing the sensors as close as possible. The moist ball of fuel was first shaped by his tongue then tipped backwards into a short flexible pipe to the fuel tank. The pipe contracted in rhythmic waves, drawing in fuel when the machine was in any position – including upside down.

Shaped like a boxing glove, the fuel tank had a capacity of little more than a litre and took almost four hours to empty – regardless of the machine's energy requirements.

As fuel poured into the tank, its thirty-five million wall jets began to spray a powerful acid, sterilising and tenderising the potentially useful, useless, or disastrous breakfast. The acid was strong enough to dissolve iron, or burn a hole in the carpet.
Too much acid would corrode the machine; not enough

and the tenant would starve. Behind the jets, hidden
motors slowly flexed the walls, churning the propellant
into a watery pulp before injecting it into the single,
eight-metre-long piston at a mouthful a minute.

The absorbent walls of his stomach and intestine were
coated with a detachable lining, replaced every two
or three days in response to the rate of damage
caused by the acid. Even if the tenant ate nothing
during that time, the machine would always be able to
extract some energy from the exhausted outline of its
own combustion system.

The intestine's first curved section contracted,
gradually opening the stomach's release valve and
drawing out the thick acid pulp. The corrosive mixture
instantly set off hidden sprinklers loaded with
concentrated detergents from the green, pear-shaped
gall bladder and the yellow pancreas. Ten critical
minutes passed before the next six metres of folds
received the neutralised pulp and began the process
of splitting it into casual blends, converting it into
usable fuel. The piston's narrow red maze was
velvet-lined with tiny brushes to stir any congestion,
maintain the flow, and soak up the fuel. For almost
two hours the raw energy had been gently extracted
through the brushes and out of the piston by a thick
tubed network linking into a single conveyor, reaching
up past the fuel tank to the machine's second largest
component — a soft red warehouse and chemical
factory. His liver weighed only three per cent of the
machine, but operating continually, it used up
almost a third of the machine's energy as it
systematically processed the fuel into more adaptable
grades for the machine to comfortably tolerate, then

stored them ready for use. It had been carefully sited to intercept and remove most poisons or drugs before they could seriously affect the machine's overall performance, continually checking the machine's temperatures and maintenance systems, holding back any surplus energy and delivering a steady revitalising supply to every component. Whenever the tenant felt sick, it was always due to a mistake in its quality control. One imperfectly assembled amino-acid link per million was enough to create nausea and there were more than 4000 million at work in just one finger-tip. Each day throughout the machine, the specialised microscopic acids made new cross references for repair and development by dividing each existing minute correlation in two. They were designed to reject spare parts that the machine had not manufactured for itself.

The cells were fragile and complicated to make. Every new pair needed the same characteristics as the original. To protect their essential connections with one vital part in the machine, streams of chemical reactions were discharged in continuous bursts to avoid their separation and immediate rejection by the system, without the crude and violent use of heat or pressure. They were constantly alert for the slightest change. Their intelligence was war by other means: the first politics of extinction.

Air bubbles and escaping gases in the piston were slowly trapped between the soft, compacting layers of used energy, gradually forcing out the water for reprocessing, turning the pulp into a dark, jelly-like humus – the first stage in the development of coal. The increases in pressure and slight rises in temperature accelerated its gradual enrichment. Pockets of marsh gas quickly developed.

The machine's exhaust was a mixture of six intoxicating gases. Highly combustible, the hydrogen and methane had increased to almost a third of the gas. When mixed with air, they could be explosive, and there was always enough in the machine to blast it to pieces. As a precaution, the sulphur in the methane added a pungent smell, like the artificial smell added to commercial gas. Without it, the tenant would probably have been unable to detect a potentially dangerous leak.

*A fart. Two.*

The machine travelled by falling forwards, backwards, or sideways. Most of its motive energy was spent in swinging its jointed legs one after the other at an anticipated invisible trail of balancing points on the ground. To increase speed, it simply fell faster.

*I almost ran.*

Surges of the thick brown liquid poured into the piston's last bloated section. The unconverted residue of a two-day-old dinner was overtaken by the fast-moving vegetable fibres of yesterday's lunch.

The thickened tube swung abruptly upwards, passed short empty cul-de-sac, crossed under the fuel tank, then looped back down in a sweep before an abrupt halt before the final short drop.

The dark raw fertiliser, containing live tomato and cucumber seeds, noisily increased speed as it became lubricated from the piston's lining. Pressure built up, compacting the waste against the water-tight exit valves. The computers instructed the upper valve to open, exposing the tenant-controlled lower valve to the full weight above.

His pleasure and relief in shitting came from the pure relaxation of every controlling muscle.

Both valves stretched wide, and the long, wetly faceted cylinders squeezed out one by one with repeated waves of compression extending over the entire piston area. The oddly sized and foul-smelling cylindrical gardens, with pointed ends from the closing valves, had been carefully scented to protect their precious contents from disturbance until the raw fertiliser could be naturally diluted with rain, and the nourished seeds germinated back in to plants. The raw materials necessary for new life.

*I pressed my fingers against relaxed, closed eyelids.*

The surfaces of the camera lenses were kept sterile by a steady flow of antiseptics ducted from the side of each lens. The protective fringed lens caps automatically swept them clear every two or three seconds, washing any germs dead or alive down tiny drainpipes at the inner corners into the air-conditioning cavities. The supply of antiseptics was topped up by two pairs of ducts beneath each eye. At the end of each nostril, the moisture finally dried in pale yellow and green crusts on the protective hair filters. Each hair was connected to a nervous system designed to detect any blockage or invasion.

*I blew my nose and produced the last remaining turd.*

The same muscles operated both actions.

His persistent running nose, its wet lining pale and soft with the flow of mucus, was continually irritated by smoking, and congested and thickened by alcohol. The overflow slid down the back of his throat to his tongue.

*I swallowed, and automatically reached for the
lavatory paper.*

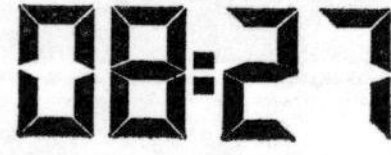

The surrounding hair had been coated in passing and was partially wiped with two thin sheets of absorbent paper. The residue would be rubbed off on the tenant's clothes during the day.

*I washed my hands and quickly dressed.*

The thick, natural hair covering that had once given
earlier tenants more flexible temperature control
and greater tolerance for seasonal extremes, had been
lost during 20,000 years of their descendants'
steady preference for hairless bodies. In turn,
everybody's self-protection had drastically
thinned, and although the tenant still had more hairs tha
a chimpanzee, clothes were a much more interesting
wrapper – adding pleasure in their removal.
The machine was now a shelter with decoration,
covered at night by a two-storeyed brick and concrete
variation.

*My wife held me close, kissed me, and hoped that I
would have a nice day.*

He was excited by the sight, smell and mere touch of her body and his memory of what her clothes concealed. Sex was like a meal — it needed an appetite to appreciate it, to be shared or eaten alone.

A screaming whine rose up from the room, louder than any Concorde, too painful to smile over. Unnoticed, he left the insane rage of the vacuum cleaner.

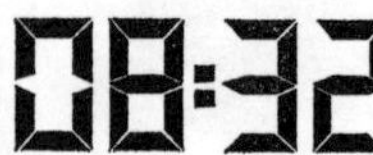

A kilometre to the West, a small shred of cloud — a thousand tonnes of water vapour — drifted into a turbulent column of air. It veered gently towards the machine, nudged by the dispersing wind into the cooler surrounding air where it quickly lost the buoyant warmth that kept it flying.

The soft tanker began to soundlessly break up as the overwhelming increases in pressure squeezed out its last moist heat. All its cooling weight was now solid water. The first drops of it started to rain down.

*I had forgotten my raincoat.*

He ran the last few steps to his car – his internal combustion umbrella. The beautifully machined compact symptom of power breathed the same air but wasted most of its energy just wheeling itself around. Its deadly exhaust was the most common pollutant. And built into every car – and every scarecrow – was a secret ambition to walk about and terrorise.

At the turn of a switch it resumed its obedient life, and smoothly accelerating, began to shrink the tenant's former time and space.

Now armoured and partly mechanised, he focussed on the road ahead, driving by precise rules that licensed him to pretend that he was in control, and, in doing so, separated him from his normal inhibitions.

On uneven ground or beginning to stumble, the walking machine with its protective complex systems of feedback and automatic correction adjusted its balance and saved itself. But as a semi-robot, a man steering a car did not inevitably come to a stop if something went wrong. He didn't get a stomach-ache if some fault in the car was developing. The machine's self-regulation in movement, which had taken millions of years to evolve, did not operate on wheels.

He only needed the edge of his mind to catch at the advancing sights and sounds, to shift gears, increase speed, swap lanes, gauge distances, and seal off the rest of reality. In doing so, the tenant blocked the machine's access to the computers' primary circuits of awareness. If they had stayed open, there would still have been some signal – some sort of clue –

which would have alerted his memory circuits to the
persistent danger. Instead, the car was operated by
his computerised, extended force of habit.
In his smallest movement there were thought patterns,
but on a level so low there was no instrument
sensitive enough to measure them.

*The dog was out into the traffic and almost under the car before I even noticed it.*

It was impossible to isolate the chief reason for any critical event, as it could not be established in any precise sense. For any coincidence, there were hundreds – thousands – of contributing causes, and no way of apportioning priorities to any of them. Anger at the dog in the road and the owner's carelessness were mixed with the fear of hitting it.

I shouted — a frozen split-second — and the dog
was gone.

His heart leapt at the computers' silent screaming
alarms. Blood pressure instantly doubled, forcing
stored energy from his liver through hundreds of miles
of elastic pipelines. Their walls tapered fast to boost
the emergency power, simultaneously shutting off
millions of tiny surface terminals, limiting any possible
superficial damage to the pale, drained skin;
restricting the supply to vital connections.
Every component, every super-tensioned joint,
instantly prepared for the impact.

*I braked and swore; my heart pounding, hands and face sweating. My mouth felt suddenly dry.*

As the shock waves eddied, the momentum of the
sudden burst of unused and now just as quickly
discarded energy swirled through the machine.
With falling pressure and restored circulation the
restricted circuits slowly reopened to normal use.
Streams of microscopic unused repair patches stuck
to the walls of his heart and its main arteries,
aimlessly building layers of resistance, constricting
the flow, forcing his heart to work harder and harder in
future crises.

The violent braking dislodged a tiny gas bubble
of digestive acid in the machine's fuel tank.
It rose slowly and erratically and, catching in the curved
neck of the tank's inlet valve, it attacked the exposed
lining, bringing an uncomfortable burning sensation.

*I belched, and switched on the radio.*

The soothing sounds were colourless and without
character. It was music that was meant for his stomach –
to be played but not heard. Digestive, gastric music.
His brain's overcharged memories repeated fragments
of the last few minutes over and over again.
Involuntary and unidentifiable recalls of fear, anger
and frustration. The shocked emotional systems
expanded the awareness of nearly killing the dog.
Everything looked different; the familiar road no
longer the same.

Twenty minutes later, he pulled in and parked beneath
a particular multi-storey mudpile – one of thousands
of almost identical blocks of steel, concrete and glass,
plastic, marble and stone. Whether transparent,
translucent or opaque, they were all merely
superheated, refined mud. To be used once and
thrown away.

The usual good mornings; easily misunderstood.
Words were a dangerous code not meaning exactly
the same to any two people. Too many words had
more than one meaning or pronunciation, yet sounded
the same. Speech had erratically developed from
three million years of groans and grunts that had
been the inevitable association with particulàr forms
of physical effort; each type of grunt emphasising a
particular type of activity.

His upright ancestors had developed an improved
valve in their combined airway and refuelling tube
to lock air more efficiently within their bellows.
Impulses from the computers regulated and modified the
amount of air passing in and out. The stimulated
valve contained a pair of highly variable elastic bands,

vibrating in the airstream.

A sound was made after the bands had been compressed together to close the valve's wedge-shaped opening, and were then forced apart by the pressure of air blown out of the bellows. After each puff they closed again. The vibrations occurred up to fifty times a second, and on more than one frequency of sound. The movement shook up and separated the air, causing minute changes in pressure before converting them to invisible sound waves. The volume and pitch of the sound were determined by the size and speed of the waves. The tonality of any word could not be established until the following note had been received. The rules of grammar were useless without the meanings of words, and the correct perception of language depended heavily upon their context and knowledge. Sounds became speech when they had been turned into notes and joined together by the computerised memory of the infinite variations of just five types of articulation.

Although there were only 3000 words stored in the computer, there was no other form of communication comparable to the tenant's ability to talk and sing.

Loudness of voice depended on the amount of air available, and the pitch of sound on the variation of tension in the bands to produce a series of different notes. The mixture that finally made a recognisable voice was determined by the individual shapes and resonances of the intervening cavities to the outside air and the position of teeth, lips and tongue.

Hearing enabled the tenant to check his own speech for volume and emphasis. He heard his own voice first in his mind, then aloud in his ears; largely through the sound being conducted directly through the structure of the machine.

Speech was meaningless energy until it could be heard, and the tenant needed only five sounds to communicate. The computers were capable of combining them in over 100,000 different ways, and there were additional inbuilt connections linked into the circuits enabling the tenant to express an infinity of sentences, aloud or to himself, making it possible to integrate to a greater or lesser degree with associates who were now more concerned with power than with pleasure or fun.

*I lit a cigarette.*

The inhaled smoke from the paper-wrapped leaves held more than 300 chemicals and solid granules. The smoke included the deadly odourless carbon monoxide, now concentrated in the machine at 400 times the safe industrial level.

The granules condensed to form a thick brown tar, lining the air ducts and forming lethal pools in the porous bellows. Microscopic smoke detectors in the air ducts set off alarms in the machine's automatic defence systems, violently compressing his lungs, coughing the smoke back out through the narrow airways at 150 metres a second.

He craved the nicotine, an oily hypnotic poison in the smoke. The concentrated poison from fifty cigarettes would be enough to destroy the machine. Half this was smoked each day, slowing and distorting his communications systems and computer activity. Small amounts gave him a sense of stimulation, then left him feeling more tired than before; so he smoked more. The larger doses became a sedative, combating and concealing his anxieties and tensions.

Thinking was the control and suppression of thoughts that existed in their own right. He sat in his multi-coloured cube staring at the plastic landscape. Nearly three more hours to sit and look elaborately at nothing. Technology was everywhere; the most powerful incentive, although the actual amount of energy that was of any benefit was very small. Machinery that could work twenty-four hours a day was only used for eight hours. Buildings were constantly being put up for fake employment.

Typewriters now reclined on the best furniture, while
people slept in slums.

For the tenant, security of employment took priority
over everything when really he needed risk
exercise to maintain his sense of self-preservation, to
stretch himself beyond his physical and mental
resources, to push himself a little beyond a point
where he was completely safe.

The machine's adaptive processes were too slow to
adjust to technology's induced rate of environmental
change. The typewriter did not care what was being
typed on it, and the mechanical computer was
indifferent to the instructions in its programming.
With no real author or responsibility, the carbon
copies and printouts had become terrifying.
The mechanical computers' mindless rules and
definitions made them capable of solving enormous
arithmetical and mathematical problems in a fraction
of a second. And they all believed that one plus one
made two. But two lamps did not provide twice the
level of illumination, and a conversation did not
double the level of sound. Trebling the temperature
did not make the machine three times as hot, and
moving at quarter speed did not save a corresponding
amount of energy.

*Squeezing my eyes shut for a moment to think, I saw beautifully changing ghostly images.*

The pressure-stimulated visual noise grew as the
randomly generated fusion of patterns in the libraries
of his eyes revealed the vocabulary of his vision.
It was a unique bound set of basic facts amongst
his drifting abstract dialect, from which the computers
assembled their visual world. He did see things
differently from anyone else.

He was so conditioned to the necessity of discipline
and the ordered world of straight lines that his brain
had difficulty in recognising and welcoming a curve.

*More and more paperwork arrived every day.*
*A particularly important letter had to be answered by*
*this afternoon.*

Both sides of the upper computer were activated as the cameras began to read. The right side issued the command for the cameras to move to the start of the line. The left side prepared to translate the incoming information.

Each half was adept at working simultaneously as they processed the verbal material. It was not essential for the cameras to *work* in unison, but for binocular vision they needed to *move* in unison.

The tenant relied heavily on the upper computer's memory store to automatically recognise and ignore familiar words, and concentrate on more interesting information. His visual perception interpreted the cameras' fragmentary and ambiguous inverted images; seeing other meanings, sensing his own version, not the evidence. If there were more than one complete and distinct interpretation, he saw most of them but not at the same time, and when he eventually understood something, he saw it in retrospect.

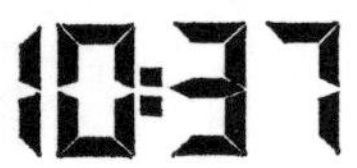

Group Investment, Expanding Production — the drift towards growth had become inescapable. If there were limits to growth in numbers, or in the size of things, then he would need to use his own computers to alter the drift. Working with only his barely tapped creative and inventive skills, he could move instead towards the options of thought and imagination rather than the sole business of producing things.

Even now he could develop an ability to plan and build on his inevitable mistakes. By becoming discovery-prone he could seize upon and take real

advantage of the unpremeditated and unforeseeable
accidents in his life.

*I sniffed and grunted at the paperwork decisions.*

The machine's natural time ran nearly fifteen
minutes behind the accepted twenty-four-hour day.
During a working day each month he was forced to
work with the machine's night-time chemical changes
as they converted fuel into energy or component use
at its lowest working temperature and efficiency.
Even the cells of the larger components had their own
time-clocks. Those connected with structural
movement had seven to fourteen-day cycles that
worked at their peak when balancing out the overall
circulation whilst the others were ticking over.
If each clock was jerked out of gear and into the same
phase, jolted by shock or sudden degeneration of the
system, they would all dangerously fluctuate up and
down together.

The heart had its own dual-circuit pacemaker clamped
around its piston walls to defend its smooth pumping
action. Independent of his brain, the pacemaker
increased the signals between the blood-pressure
controls and the heart's flow capacity, operating the
inlet and outlet valve with short electrical impulses.
As pumping increased, pressure rose against the
outlet valves. Any significant increase was transmitted
up to the computers, which then sent over-riding
instructions to the pacemaker to slow down. As the
pump's activity slowed, the fall-off in pressure
reduced the transmission to the computers. The effect
of the signals on the rate of circulation was immediate
and short-term; essential for the machine's survival
in a world of suddenly changing demands.

In addition, there was a slower and longer-lasting
method for controlling the circulation's volume and
concentration, by monitoring the computer's
instructions with the release of chemical messages
injected into the tidal flow. The pressure-operated cells
within the pump itself could then be activated.
The overlapping safety systems combined into a
sensitive mechanism influencing the machine's
ability to respond to internal and external crises
quickly and efficiently.

The controls did not rely on external events.
The tenant could operate them from his own card
index of ideas, emotions and fantasies.

When the machine was in good condition, the pump
could not be stressed beyond its reserve powers of
operation. Working flat out, the muscular motors
would reach their limits well before the pump was
damaged. However, if its valves were damaged or
feed lines blocked, the pump's reserve capacities
could be exceeded by even the smallest exertion.

Emotional stress automatically released energy
propellants stored beneath the machine's skin for
the sudden anticipated response to danger.
Sitting comfortably in his office, the tenant did not
need to move, although he was unable to prevent his
feelings being associated with the machine's inbuilt
sense of danger. His blood pressure substantially
increased, placing more strain on his heart than if he
had been crossing the road — where circulation was
assisted by the muscles involved in walking.

*The telephone rang.*

Face to face with the receiver, the familiar tensions returned of not being able to hear a conversation with both ears.

Normally, when sound came from anywhere other than straight ahead or behind, its intensity differed as each ear was at a slightly different distance from the source. Sound waves travelled relatively slowly and there could be a time difference in reception of almost a thousandth of a second. The machine could easily detect this, locate the source, and enable the sound to be heard in stereo. The stored unnatural memories of previous telephone calls generated an artificial response to the electronic words.

The computers were very sensitive to the real time, keeping track of the date, knowing when to refuel or rest or be alert. Instead of asking himself, the tenant chose to wear a watch to check the hours and minutes.

*Lunchtime.*

He only understood the space he occupied when he moved through it. The quality of his vision was essential to the machine's ability to balance and stimulate his customary indifference to his surroundings. His changing perspective was recorded by the computers as they located the machine's relative movement, adjusted its speed and position, and simultaneously rechecked the amount of space remaining and the obstacles in his path.

*Outside, the surprise of a warm, sunny day.*

The machine had no effective self-protection against
sunlight, although the danger had always existed.
Short periods of exposure activated the machine's
defences, partially shielding the skin by causing its
deeper brown colouring to rise to the surface,
trapping some of the incoming radiation before it
could cause damage.

Sunlight also stimulated the manufacture of vitamin D
beneath the surface of the machine to maintain its
healthy bone structure, as its thermostats automatical
adjusted to the changing intensities of sunlight and
shadow. Too much heat triggered temperature-reduc
instructions, drawing in more air and expanding the
circulation area immediately beneath its surface for
the excess heat to be transferred more quickly to the
skin's microscopic water-filled cells that constantly
drained on to the surface.

With the machine's movement, or merely in a warm
atmosphere, the cells' output was raised to keep down
the overall temperature, evaporating up to two
litres of water a day. More than half filtered out
through his clothes. If the machine's surface cooled
too fast, reverse instructions reduced the loss of
internal warmth. Minute strands of smooth muscle
attached to each hair root contracted, pulling the
hair upright, trapping his body warmth and creating a
heat-insulating layer.

*I shivered as I crossed the road.*

His skin paled as it compressed further to reduce perimeter circulation and heat loss.

Some of the clean fresh air had been around for a long while. Every time the machine drew in air it was likely to suck in at least thirty million molecules exhaled by Dickens, Napoleon or Leonardo da Vinci.
The electrical condition of the air had a profound effect on how the tenant felt. Invisible electrically charged particles – ions – from the negatively charged Earth continually flowed upwards through him, increasing the beneficial flow of the machine's own electricity, finally collecting in the atmosphere to be returned to Earth as lightning. Negative ions helped the machine to absorb the water in its components, affecting illness, nutrition, and age. The tenant's tensions and depressions were frequently due to the increase of positive ions given off by buildings and their electrical equipment.

*The adjoining restaurant chairs were only a few millimetres from mine, where friends confided to each other the fascinations of drinking and driving, or their indecision about the merits of double glazing. I glanced at the menu.*

The Sun was the machine's only source of fuel,
although only green plants could use its energy
without additional conversion. By building
themselves with sunlight, silently and effortlessly
converting light into chemical energy, they provided
the particular characteristics of air needed by the
machine for its own processes of energy conversion.

The machine could refuel from plants, or more
efficiently, from animals who were plant eaters.
All meat was grass. The machine preferred not to rely
on plants as it used up too much of its valuable energy
converting their bulk into usable fuel. A lettuce was
little more than one per cent efficient at transferring
energy, but a cow was ten times better. The vital factor
in any conversion was the initial cost of saving the
energy. Although the tenant might not see the Sun for
several days, for a patch of grass to provide the machine
with enough fuel it needed 3000 kilowatts of sunlight
every day.

*Sausages and baked beans.*

The lack of fibre in the highly refined lunch meant that
it was eaten twice as fast and took twice as long to be
processed through the machine as it should have.
The two pig fingers fried to a crisp brown in a spoonful
of orange-coloured seeds represented the rapid and
largely unchecked evolution of pre-packaged fuel
production techniques. They broke down the fuel's
natural chemical structures, modifying and
rearranging them in complex and ingenious ways.
If he took in a lot of garbage food, he would eventually
look like a garbage truck. Each day he had to
coexist with almost 2000 unfamiliar chemicals
arbitrarily added to his diet, as the machine operated
with the power supplied from the synthetic
nutritional building blocks. The adjustment was
further complicated by the interaction of the chemical
additives and natural fuel components, and by the
combined fuel and the machine's methods of
absorption. The natural chemicals in the mustard – a
relic from the eighteenth century to sterilise suspect
food in the tropics – destroyed all the bacteria in his
stomach, good and bad together.

Fuel energy was measured and distributed to the
motor circuits and to the renewal and repair
mechanisms. Some was stored, but more than half
the energy was transferred into heat to maintain the
machine's working temperatures. Absorbing and
digesting the fuel required the diversion of part of the
blood supply from his brain to his stomach and
intestines, making the tenant feel tired, overcoming
any real inclination to re-engage in heat-generating
movements, such as work.

Four hours after refuelling the weight of the machine

had not increased by the precise weight of the fuel
due to moisture loss through its skin, the loss of
energy during combustion, and from the saturated air
passing out of his lungs.

He quickly wasted sixteen litres of water to flush away
another half-litre of urine, pausing on the way
back to wash his hands with liquid soap and
hygienically dry them in a stream of hot air.
He combed his hair, adjusted his tie and shook hands
with a colleague. Nearly half his illnesses were caused
by touching something or someone. Almost 200
different types of virus were exchanged during the
handshake, surviving for more than three hours on the
warm, moist palms of his hands.

At the conference, he felt like the others around the
table. Trapped in their computerised suits and
graph-paper shirts. He appeared little different from
the previous year, older perhaps, carefully trying not to
betray the inevitability of the machine's cell
degeneration. Being close to others had become a
tactical weapon. His little refuge of privacy and
anticipated separation was under attack. He could
discreetly change its size, depending upon the
necessities of the moment. He could endure the
inadvertent physical pressure of someone in the
crowded meeting, but not if he was alone on a wide
and empty street. The fixed stare at a younger
colleague was a conscious penetration out of the
shelter; a message that carried his interests in sex, age,
race, money and influence. Some form of separation –
almost a barrier – was essential.

The machine's ingenious protective mechanism of

hearing could not cope with the sudden and repeated
high frequency sounds. Hearing loss was caused by
his automated environment – the constant
background noise of machinery, the surrounding
babble of conversation – and the loss was
irreversible. High-frequency noise produced more
errors and the symptoms of nervous tension.
His headaches, insomnia and indigestion now relied
on artificial rescues and his ingenuity in first elating
then sedating, expanding then contracting,
suppressing or blocking the machine's natural
rhythms.

He chain-smoked in reaction to the surrounding
tensions. Sedatives in the smoke added conflicting
signals to the chemicals in the coffee. Tense and
relax, tense and relax. His hands trembled as the
opposing instructions shook the machine in invisible
disquiet at the tenant's unrelenting search for the
chemical control of his mind. Sudden relieving
laughter broke out at an unexpected joke.
Laughter was a smiling cough that started in the
stomach and ended at the eyes. The machine's piston
was unprepared for the abrupt agitation. Bubbles of
escaping gas were dislodged and stirred on downwar

*I held on tight to the disconcerting fart.*

He had quickly learned that to reach into most facets
of the society in which he lived he needed the social
approval of his clearly demonstrated control over
his fear and apprehension.

To compensate, he had adapted part of his circuits of
abstract thought to take in his increasing fantasies.
Without these short cuts, the computers would have
continually searched their memories for sensory
inputs — interpreted by the tenant as life being one
damn thing after another. The computers normally
prevented anything from activating his pleasure
circuits for very long, and to keep these circuits
alive and dominant, the tenant had to instruct them
to find more and more sensory pleasures.
As the computers submitted, the machine's natural
energy no longer built up in a normal way, and its
natural self-regulation and release did not take place.

The tenant's changing use of the machine left him
exhausted. Although back at his desk he was busily
moving around, it was often repetitive motion that
would not revitalise the machine. Circulation became
sluggish, eddying into pools within its components.
Instead of pumping the fresh supplies of energy
around faster, fuel penetration slowed. Instead of his
muscles becoming more efficient by pushing each
other slightly beyond their previous output, they were
becoming increasingly ineffectual, incapable of
reaching even their own former ability.

Exercise would have burnt out the dangerous blocks
of stored solidified energy that were now building up
in soft, insulating layers of fat beneath his skin.
With the increase in weight came additional pressures

and strains. The layers of fat were supported by a strong fibrous network. If the layers became thick and heavy, they would bulge out between the fibres, distorting the contours of the machine. Some of the fibres would change into fat as the tenant felt himself under stress and the machine sought to protect itself against the tenant's anxieties and depressions.

Poor circulation caused the fat to stagnate, preventing the machine from quickly using its potential energy when the tenant felt better.

The machine's surface was its largest component. It contained an extensive texture of touch sensors located immediately beneath the surface and deeper inside the skin. The machine had nine senses, although the tenant was only aware of five – sight, taste, smell, hearing and touch. Touch involved five senses: heat, cold, position, pressure and pain. Memory of touch improved with successive events until it had eventually become the tenant's dominant sense of reality. The machine's ability to feel shapes and textures provided the computers with more precise information about external events than sight or hearing.

The delicate sensors of pain and temperature fed the machine with subtle degrees of information, for the machine to alert itself automatically and react in self-protection, often before the tenant was aware of them. The machine relied on touch more than anything else. Often it was not until an object could be felt that the computers were finally convinced of the existence of anything.

Each type of sensor formed minute electrical fields
sensitive to a wide range of stimulants. Receptive
fields overlapped with others, and when pressure was
applied to a particular part of the machine, several field
were activated at the same time. Reception varied in
intensity with the concentration of sensors and the
changing relationships between the fields. The rate of
signals and distribution of impulses instigated by a
caress, or scratch, or tickle were initially grouped as
moving itches, and their collective effect travelled
along the same fibres to the computers. The individual
messages were then separately decoded from the
combined signal pattern and filed as particular
sensations.

Sensations were rapid dreams. The tenant's awareness
of what was happening to the machine relied upon
his interest in the upper computer's overview of
sensory information. The regulations of awareness
were clear: if he did not know what he had been
doing, how could he decide what to do next?
By withdrawing from his feelings of insecurity, he
could then distract himself with meaningless activities
regularly fidgeting, or scratching himself, or chewing,
or fiddling with anything that was available.

Patches of the machine's surface were constantly
rubbed away, and were either trapped in the tenant's
clothing or drifted into the air. An entire outer coating o
the machine was shed every month. Up to eighty-five
per cent of the surrounding dust was shed human skin,
and there was at least one complete human skin
floating constantly around the busy office.

Within half an hour the tenant had rubbed his eyes,
scratched his head and rewound his watch, shifting
his position in more than 2000 tiny ways. His hidden
anxieties grew with his boredom and frustration.
A number of touch sensors short-circuited, anticipating
the phantom rescue signals of distraction as the
masturbatory need to itch and scratch increased.

Sex was the tenant's most enjoyable and oldest
distraction. He had found that it was all a matter of
understanding that it was not only people that grew
up: fantasies did as well. When he had found himself
with a full-size body, he had needed fantasies to go
with it, different from the ones he had had when he
was a child. The fantasies about sex that he had started
with had become confused about what was the
cause and what was the effect.

His wishful thinking about the attractive girl in the
photo-copying room was heightened by the clothes
she wore and the way she moved inside them–
especially the clothes that he could freely imagine she
wore beneath the visible, tantalising outer layer.
It was only Adam who had said to God, 'I was afraid
because I was naked.' For Eve, modesty had already
become an attitude, a means of becoming more
desirable.

The first clothes were not put on out of a sense of
shame but as protection against natural dangers.
It was later when the areas that were concealed and
seen less often naturally became more attractive.
More and more importance was attached to what
was underneath, and that whatever was being
covered was also on show.

The tenant's own clothes concealed his convenient penis. Its hydraulic system could be operated manually or by remote control to provide his most personal and pleasurable relaxation. Hidden in its cushioned, cone-shaped tip was a concentrated ring of touch sensors. Their millions of micro-circuits of pleasure remained inactive behind a flexible protective cover until the tube began to extend. Within five seconds, it could double or treble in size; transforming the soft and pliable shape into upward-curving rigidity, swiftly rising with increased hydraulic pressure against all the laws of gravity. Extending, the protective cover would draw back, exposing the stimulated circuits activated by the sudden release of the tube's constricting dimensions. The energised sensors would then react to the slightest touch or sensation. The computers' overview of sensory information would then be rapidly and completely swamped by the tenant's highly charged memory circuits of sexual fantasies, and his dreams of complete involvement. While they lasted, they formed a dependable screen and safety valve against his continual pressures of anxiety and tension.

The tiny oval trigger mechanism for the circuits was located deep in his brain, between the computers. Internal or external events could over-ride the tenant's firm control of his mind, linking both computers into a single urgent process of stimulation, fully absorbing both tenant and machine.

With any sudden increase in circulation through the inflatable linings, muscles controlling the return flow at his penis's junction with his body simultaneously contracted. The inflation would continue by the

sustained inlet pressure from the machine's circulation
until the tube reached its maximum extent, jutting up
hard out of the machine. A hot, pungent, purple-
headed daydream.

Behind and beneath the tube was suspended a flexible
bag, divided right and left by a fibrous partition.
In each compartment lay two highly sensitive grey
oval balls, the size of chestnuts. The bag and partition
were lined with a smooth film to enable each ball or
the entire assembly to move independently, or retreat
back up into the machine with approaching danger.
Inside the grey exterior 1000 million descendants
of the machine were manufactured every day, each
one taking ten weeks to make its unique
microscopic motorised intelligence that would need
to grow 70,000 times its length to reach full size
and increase its weight by 10,000 million.
Their construction was programmed by a long coiled
string of a four-letter chemical alphabet, where the
codes for building anything from a germ to an
elephant could be spelt out. The monitoring process
that checked the start and stop signals for reading
the code was the basis of life.

*I forced myself back to work, work of national
importance. A warm, tickling sensation built up
in my nose.*

The machine suddenly reacted to the air-conditioning's warning system of the invading viruses. When the sensitive linings were attacked, messages to the computers returned as signals for the bellows to expand and then suddenly contract. Air was blasted out with the irritants in an explosive rush, faster than the speed of sound. Unless the sneeze was immediately trapped, minute droplets of the moist discharge containing hundreds of germs remained suspended in the air for hours, to attack other machines.

Stifling the sneeze drove the irritants into adjoining channels, squeezing in through the linings to surface deep in the machine's circulation.

For the viruses to be entirely passive depended upon any of the machine's sixteen billion blood cells having a conveniently free surface-receptor for the viruses to attach themselves, induce the unwary cell to swallow them as food, and become infected. There were only a limited number of receptors to each cell and the viruses attempted to attach themselves to one particular receptor, making the cell impervious to other viruses as they competed for space on its surface. Once inside, the viruses quickly developed their own special techniques for escaping undue attention from the machine's defence systems by ingenious methods of deception and disguise. They needed to prolong the life of the natural cells to spread and colonise by debilitation and destruction. The machine counteracted by stepping up the concentration and variety of its defences to discover and destroy the invaders.

Every cell contained thousands of microscopic hereditary blueprints to determine whether it would succumb to the invasion, or the form and degree of immunity response. In turn, they were in constant communication with a slowly circulating parallel network of defence. The transparent lymphatic network had filter-glands strategically sited around the machine to trap the viruses, dust, or other dangerous particles before they could reach the machine's vital components. Each filter made its own defence cells to increase the machine's immunity and destroy the invaders or neutralise their effects. Together with the patrolling cells manufactured in the cores of his bones, the filters released streams of chemical adhesives to bond the elusive viruses together, making them more apparent. As soon as they had been located and identified, they were attacked by the defending cells who swallowed them. Then to prevent themselves becoming a potential danger to the machine, each one was fitted with a self-destructive mechanism timed to activate after a day's operation.

The tenant needed a completely new approach to his inability to cope with the machine's most frequent malfunctions, as medical knowledge had no cure for even the common cold and the cost of medical care was rising so fast that no country, not even the wealthiest, would soon be able to afford it. Having identified and almost eliminated the fatal, infectious diseases, medical knowledge was helpless against the rest. It often inspired or prolonged suffering to no useful effect, obscuring the causes of new illnesses — stress, pollution, and artificial diet.

Modern medicine had become a major threat to health in its claims for the elimination or delay of pain, suffering and death. It had become normal for the tenant whenever he experienced any physical or mental discomfort to turn to medicine for a magic cure, instead of using his own natural healing abilities to reshape his existence. The populations of advanced countries now had a lower expectation of life despite the years of expensive check-ups and the enormous hospitals.

*I needed a drink.*

The carefully planned evening mass escape was at
its peak.

The tenant had spent nearly a fifth of his life in some
way or other with a car — driving it, parking it, cleaning
it, and earning enough to run it. In return it had
increased his daily average speed to six kilometres an
hour. Apart from fresh food, nothing devalued faster —
yet cost per kilometre it was as economical to run as
a ballpoint pen.

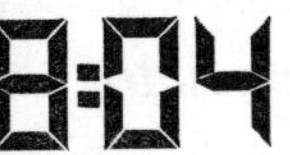


Home again. The small, embracing reassurances of
every room greeted the tenant with his own colours
and sounds. His wife's welcoming tender kiss was
heightened by the machine's need to increase its
immunity by regularly exercising with other known
viruses.

The over-ruled circuits relaxed. The tenant's feelings
of wanting to break out of his body slowly subsided.
The elaborate fantasies that he had needed during the
day to force the machine to complete what he had
started, faded away. He loosened the carefully
knotted woven length of Japanese worm excrement
from around his neck, kicked off the highly polished
remains of peace-loving cows from his feet, and sipped
a clear concentrated distilled extract of maize, malt and
rye flavoured with juniper berries. The drink was
carefully diluted with carbon dioxide, fruit acid, sodium
citrate, saccharin, quinine — to resist malaria — a thin
slice of lemon, and a cube of frozen water.

The second gin and tonic expanded his feelings of
relaxation. With the additional loss of inhibition came
the need to communicate.

*We talked about our day as I ran the water for a bath.*

Naked and almost totally immersed in the soothing soapy water, the tenant relaxed. Buoyancy reduced the weight of the machine by almost a third. The fully acclimatised bacteria in the moist tropical forests of his armpits remained undisturbed. The population of each vegetable republic rarely fell below 100 million, and no amount of scrubbing would remove them.

There were more living organisms on the surface of the machine than there are people alive on Earth. Although its naturally salty and acid landscape was constantly breaking away, making a difficult shelter for bacteria which had not been specifically developed to live there, the heat and moisture in the bathroom caused the larger colonies to expand and break down into smaller groups, doubling their number during the next ten hours.

With relaxation, the machine's exit valves slowly opened, their contents gently restrained by the pressure of water. The hollow scar in the centre of his stomach slowly subsided. His navel was a lifelong reminder that he had originally been a parasite attached by a bypass fuel line, whilst the independent machine had taken its full nine months to evolve.

After twenty minutes the machine's skin became hot and pink, soggy and vulnerable to infection.

He stood up and towelled away a quarter of a million bacteria. They floated into the air on their tiny carpets of fertile shed skin. Dressed in clean clothes he began to feel hungry again.

Delicious smells from the roasting carcass of a

plucked, thawed, and stuffed domesticated dinosaur wafted in from the kitchen. They sat and ate slices of the chicken's pale succulent muscles with the boiled roots of cultivated weeds, between mouthfuls of French red fermented grape juice. The subtly aromatic juice increased their interest in the fuel – its dilute acid destroying the remaining doubtful bacteria that had survived the intense heat. In the machine's refuelling system the fermented grape juice improved the overall quality of the fuel and added small amounts of chemicals essential to growth and repair.

The distilled alcohol in the glass of cognac circulated around the machine, supplying muscular energy and an anaesthetic to the communication systems. The anaesthetic slowed the machine down and impaired its efficiency. It stimulated his kidneys into losing more water than they were taking in, dehydrating the machine. His liver was temporarily put out of action. Coffee reversed these effects – made possible by a handful of beans frozen in the snow, which the defenders of Vienna had discovered after the Turks had withdrawn.

He relaxed in front of a flattened glowing fluorescent screen divided into 625 horizontal picture slices. Each slice was transmitted in sequence making a complete moving picture of over a million coloured dots, linked to a sound track. He interpreted this artificial world of electronic projection as a more fascinating extension of his own. He only saw and heard what he was meant to, in brighter, clearer sounds and colours. Regular viewing had softened his critical awareness, the tiny flickering dots

continually failing to become proper signals of
sensory information.

Undressed and in bed he switched off the lights and
closed his eyes to reality, but then he was too tired to
prevent his memory remembering today's events.
Every recollection was as clear and sharp as the
original: if he wasn't part of a solution then he must
be part of a problem.

**23:07**

The nightly tranquilliser, capsule or cognac, was
designed to blot out the computers' natural energy
conversion and switch off the internal electricity to
his over-stimulated brain. In doing so, it encouraged
the tiredness and apathy that it was meant to combat,
withdrawing him further from his natural stimulants
of food, sleep and social contact. He knew and cared
more about the oil in his car than about the possible
side-effects from his tranquillisers.
Enkephalin, an anaesthetic — made by his own brain,
three times stronger than morphine — remained
unused. He could not recall the words of his own
prescription. He could not recognise the correct
place to start and stop reading, unable to exploit the
natural machinery in his own laboratories.

Instead, he used his dreams as long-running plays to
act out the secret anxieties and ambitions that he was
unwilling to consciously acknowledge. His dreams
were as influential as actions, their representations
as real as tables and chairs. He turned over, pressing
himself deeper into the warm bed.

**23:20**

Millions of dust mites had waited patiently all day.
They began to safely graze on the flakes of dead skin

which filtered down through the sheets to them
every night.

The computers needed hardly any energy to function,
remaining alert even when he was deeply asleep.
Information poured through their circuits at the rate
of ten 500-page novels every second, working harder
now than during the day. Throughout the night, they
alternated between two distinct stages of maintenance
and repair: from the tenant's deeper and deeper levels
of quiet and relaxation when the machine carried out
its most delicate repairs, to the periodic bursts of
revitalising energy in circuits blocked by the tenant
during the day. These dramatic changes in the
computers' electrical behaviour increased circulation
and air intake. Behind their shuttered lenses, rapid
camera movements followed the vivid dreams
seemingly released from the memory circuits at
random. The dreams were caused by the computers
as they sorted the events of the previous day before
final storage. The tenant's thoughts were constantly
shifted to recall an accurate memory map and check
that it was correctly filed. The connections between
the prospecting memories demanded an enriched
language to reassemble or replace the faulty patterns.

Each memory cell had 10,000 connections with other
cells, and they all needed to talk to each other for the
correct contacts to be made and maintained.
For his memory to be effective, only the right piece of
memory could be selected and stored in its exact
sequence. However, the connections could be linked
by misfiled information, or where the direct links had
failed they could be supported by fibres spreading
in from adjoining connections.

The number of temporary circuits was not merely
necessary for the message to be efficiently transmitted,
but was duplicated or triplicated in case of blockage
or breakdown. The message selected the easiest
route, and despite the amount of signal traffic and
possible diversions, there was virtually no
interference between signals. When the correct
information was accurately packaged and properly
stored, it operated to the exclusion of all other
memories on the principle of selective competition.
The communication fibres linking the billions of cells
in the computers lacked the equipment for repair
or replacement. Elsewhere in his body, when the
connections or insulation were damaged, they were
automatically dismantled and replaced. The new fibres
grew within the outlines of the original routes, building
up the damaged networks until they became fully
operational. However, once a connection was
damaged or broken in the computers it remained
useless for ever.

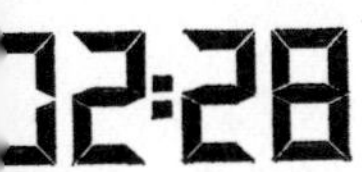


The computers' night-time electrical wave patterns
gradually became larger and slower as the tenant
receded further from the world, sinking slowly to his
lowest resistance, the machine pulling him on down

with slower and slower circulation. Operational temperatures drifted to a dangerous level, hovering fractionally above a point beyond which he could not recover from his complete loneliness and fear.
A careless or unforeseeable change in the machine's communication links, and he would begin to die.

Simultaneously, the machine's most fragile components reached their peak of activity in injecting their minute chemical messengers into the drifting circulation. Their combined action and interaction were delicately balanced to determine the rate at which the machine could develop its repair mechanisms, and the tenant to combat his apathy and depression. The circulation had to be at its lowest ebb for the messengers to survive intact as they were swept along, kilometre after kilometre. They had just twenty minutes to reach their destinations before the effect of the first arrivals began to restore circulation and destroy their intricate operation.

The tenant's dominant sensations of travelling culminated in turbulent multi-coloured dreams of flying — his anxieties about death changed to defying gravity, with its association of sex. The rolling landscape focussed on the swirling maze of woods and trees and and on down to the soft contours of ripe fruit, flowing one into another.

*I stretched out my hand.*

He felt her softly modified sweat glands. In each full
curve their twenty radial ducts converged to open
separately into a highly sensitive pink nipple.
They stiffened. So did his in useless flat imitation.
The excitement of dreamily touching everything that
they both liked, as often as they liked, erased the last
tiny barriers. At first soothing, then urgently, he felt
the growing pounding heave, his stomach turning,
and tingling nerve ends.

*My head felt light, and the rest of me felt warm and all in one piece.*

The machine reacted faster than the tenant.
Long before he even recognised his own gentle
emotive words, the machine's pattern of electrical
activity radically changed, darkening its surface with
the increased flow of circulation. When clothed,
blushes were normally confined to his face, but now
they spread across his entire naked body. Fantasies
changed to realities and back again. Night and day-
dreams became overwhelming fact, only more so.
A complete involvement with both focussed his
awareness, as his penis magically swung up into an
urgent rigid arc. 200 million microscopic descendants
began to move. The need to live beyond every
constraint dominated the entire machine. His balls
doubled in size and began to rise independently; the
right rising slowly and steadily, whilst the left
alternated up and down before finally ascending.

The swelling of the folded rim between her legs
reduced its entrance to half its normal size to grip
him. Its colour changed from pink to bright red as its
humid petals opened. They collided gently together.
Apart from using his tongue and fingers, it was the only
way that he could enter another machine. Each had a
padded protective bridge to soften the impact.
Beneath the padding, tough structural fibres were
laid down in lines which exactly corresponded to
resist the collision. The hollow interior was set out in
criss-cross patterns to brace the inside, making them far
stronger than they ought to have been. Both machines
began to lose weight as their temperatures
soared and the moisture loss through their skins
increased. Circulation, flushing through at 200 litres
an hour, doubled, then trebled. The machines
rhythmically swung apart and together again, faster

and harder. Each machine reached a total tension, an
ecstasy, which suddenly spilt over in a frenzy of
involuntary muscular spasms as they both automatically
reacted to save themselves from destruction.

Ninety million potential architects, fifty million
furniture designers, twenty million artists, doctors and
writers – more than were alive on Earth – all began to
move. Six million carried the blueprints of cancer.
They all had brown wavy hair, brown eyes and strong
teeth. Twenty million were smaller and female.
They had been initially kept in a coiled six-metre-long
cluster of tubes in his balls, before being stored deep in
the machine. Thirty billion descendants were made each
month and if not used they were automatically
broken down and remade. Production could only
occur when their temperature was three degrees
below that of the machine; higher temperatures both
prevented their production and killed those in storage.
Lower temperatures only slowed manufacture.
Carefully frozen, they would remain alive for years.
Ice was the only aphrodisiac.

The descendants had been immersed in a
protective white fluid to keep them together.
As their surrounding muscles relaxed they poured into
his penis.

*I cried out to break the tension.*

Violent contractions along the tube's length fired
all 200 million deep into her. Four high-pressure bursts
at one second intervals were followed by weaker,
irregular charges. The tube withdrew, deflating almost
completely in the time it had taken to rise.

Inside her soaking, hot and recently violent darkness,
most of the descendants missed the narrow exit at the
far end, moving aimlessly around to be attacked and
destroyed by evening.

Fifty million found the opening and travelled deeper
in, until the exit smoothly widened then separated
left and right. Once a month, from the far end of one
of them would burst the other half of the equation.
Compared to one of his sperms, it was as big as an
oval six-storey building which the last remaining
fourteen million would try to break and enter.
Usually only one sperm could succeed before the
walls became impenetrably hard.

16:45

The near-exhausted machines slowed to a halt.
Their automatic controls gradually took over,
adjusting circulations, air intakes, pumping and
temperature levels. Sudden relaxation caused
dizziness and nausea.

The tenant no longer felt that he was a particular size
or weight, or that time existed. He only consisted of
his mind, self-existent in natural timelessness.
It was easy to believe now that the world was
controlled by the supernatural, where religion fluctuated
somewhere between magic and science, and could be
influenced by his thoughts. Which had come first – the
magic or the science? The egg or the hen? It all made

sense. A hen was an egg's way of making another egg. Still asleep, he started to laugh. It was so logical, and he could not dream why, but he knew that he was near waking when he dreamed that he was dreaming.

A car started in the street.

*I was awake. Conscious. Alive. My new erection — a pleasant memory — or was the constriction just to stop me peeing?*

*Another Friday. The only difference now, the only thing that really mattered, was that today was the first day of the rest of my life.*

# Bibliography


Andrews, M. L. A. *The Life that Lives on Man*, Faber & Faber,
London, 1976

Beluzzi, J. 'The Brain's Own Painkiller', *Nature*, 15 April 1976
Berland, T. & Spellberg, M. A. *Living with your Ulcer*,
St James's Press, London, 1971
Brewis, R. A. C. *Lecture Notes on Respiratory Diseases*,
B. H. Blackwell Ltd, London, 1976
Burkitt, D. P. 'Diet – Diseases Linked by Lack of Fibre',
*Nursing Times*, London, 8 January 1976

Cairney, John & J. *The Human Body*, Peryer, Christchurch, 1973
Chanarin, I., Brozovic, M., Tidmarsh E. & Waters, D. A. W.
*Blood and Its Diseases*, Churchill Livingstone, London, 1971
Clanose Publishers. *An Index of Possibilities – Energy & Power*,
Clanose/Wildwood House/Arrow Books, London, 1974
Cooper, I. S. *The Victim is Always the Same*, Harper & Row,
New York, 1973

Diagram Visual Information Ltd. *Man's Body*, Paddington
Press, London, 1976
Draper, I. T. *Lecture Notes on Neurology*, B. H. Blackwell Ltd,
London, 1976

Feldstein, M. S. 'The Medical Economy', *Scientific American*,
New York, September 1973
Fleming, J. S. & Brainbridge, M. V. *Lecture Notes on
Cardiology*, B. H. Blackwell Ltd, London, 1977
Foxen, E. H. M. *Lecture Notes on the Diseases of Ear, Nose &
Throat*, B. H. Blackwell Ltd, London, 1975
Freese, A. *Pain*, G. P. Putnam's & Sons, London, 1974

Gomez, J. *A Dictionary of Symptoms*, Paladin, St Albans, 1975
Gomez, J. *How Not to Die Young*, Pan Books, London, 1976
Gordon, J. E. *The New Science of Strong Materials*, Penguin
Books, London, 1975

Gottlieb, S. 'Mental Stress & Apathy – The Western Disease', *The Observer*, London, 19 January 1975
Gould, D. 'Killer Diseases', *The Daily Telegraph*, London, 20 February 1976

Hall, R. 'Is Nutrition a Stagnating Science?', *New Scientist*, London, 2 January 1975
Harris, R. J. C. *Cancer*, Penguin Books, London, 1976
Hope A. 'Listening Through Bones', *New Scientist*, London, 17 October 1974
Hotton III, N. *The Evidence of Evolution*, Penguin Books, London, 1973

Illich, I. *Medical Nemesis*, Calder & Boyars, London, 1974

Jarvis, D. C. *Folk Medicine*, Fawcett Crest, Connecticut, 1958

Lang, T. *The Difference Between a Man and a Woman*, Sphere Books, London, 1973
Leniham, J. *Human Engineering*, Weidenfeld & Nicolson, London, 1974
Lewin, R. *The Lives of a Cell*, Garnstone Press, New York, 1975
Lewis, P. & Rubinstein, D. *The Human Body*, Hamlyn, London, 1975
Lishman, R. & Lee, D. 'Vision in Movement and Balance', *New Scientist*, London, 9 January 1975
Lock, S. & Smith, T. *The Medical Risks of Life*, Pan Books, London, 1976
Lonberg-Holm, K. 'Virology – Susceptibility to Infection', *Nature*, 26 February 1976
Lowen, A. *Love and Orgasm*, Signet, London, 1975
Luce, G. G. *Body Time*, Sun, Melbourne, 1971

Mackarness, R. *Not all in the Mind*, Pan Books, London, 1976
Marais, Eugene *The Soul of the White Ant*, Anthony Blond, London, 1971
Mellinkoff, S. M. 'Chemical Intervention', *Scientific American*, New York, September 1973
Melzack, R. *The Puzzle of Pain*, Penguin Books, London, 1973
Memler, R. L. & Wood, D. L. *Structure & Function of the Human Body*, Lippincott, Philadelphia, 1970
Mills, I. 'Can the Human Brain Cope?', *New Scientist*, London, October 1975
Moyle, A. *Conquering Constipation*, Thorsons, Wellingborough, 1976

Pauling, L. *Vitamin C and The Common Cold*, Pan/Ballantine, London, 1972
Pugsley, A. 'Keeping an Eye on Technology', *New Scientist*, London, 11 July 1974

Reid, G. C., Isaksen, I. S. A., Holzer, E. C. & Crutzen, P. J.
'How Geomagnetic Reversals May Wipe out Animals',
*New Scientist*, London, 29 January 1976
Rodale, R. *The Best Health Ideas I Know*, Rodale Press,
New York, 1974
Rycroft, C. *Anxiety and Neurosis*, Penguin Books, London, 1976

Sanger, F. 'Milestones in Genetics', *Nature*, 24 February 1977
Schachter, S. & Lewin, R. 'Man as Mechanism', *New Scientist*,
London, April 1975
Science of Life Books. *The Common Cold*, Thorsons,
Wellingborough, 1972
Scott, C. *Sleeplessness*, Athene Publishers,
Wellingborough, 1969
Sinsheimer, R. 'Troubled Dawn for Genetic Engineering',
*New Scientist*, London, October 1975
Smith, T. 'Why Your Body Needs Those Golden Slumbers',
*The Times*, London, 13 February 1976
Stanway, A. *Taking the Rough with the Smooth*, Pan Books,
London, 1976

Thomas, L. *The Lives of a Cell*, Garnstone Press,
New York, 1975
Tucker, A. *The Toxic Metals*, Pan/Ballantine, London, 1976

Ubell, E. *Live Longer Live Better*, Victor Gollancz,
London, 1973

Warin, J. F. & Ironside, A. G. *Lecture Notes on the Infectious
Diseases*, B. H. Blackwell Ltd, London, 1975
Wright, H. B. & Bailey, A. *A Longer Life*, BUPA Medical Centre,
Blackie, London, 1976

# Index